AF737482

LICE

FIG. 1. The Body Louse, *Pediculus humanus* Linn. ×35.
(Photographed from a model in the Insect Gallery) [BMNH Neg. No. 50508].

BRITISH MUSEUM (NATURAL HISTORY)
Economic Series No. 2A

LICE

Fourth Edition

TRUSTEES OF
THE BRITISH MUSEUM (NATURAL HISTORY)
LONDON: 1969

First Printed 1942
Second Edition 1948
Third Edition 1954
Fourth Edition 1969

MADE IN GREAT BRITAIN AT THE PITMAN PRESS, BATH

LICE

THE Lice affecting man are called by a variety of common names such as "Cooties", "Grey-backs", "Crabs" and so on. Together with a number of other noxious creatures they also come under the wide designation of "Vermin".

Lice found on man belong to an order of insects, the Anoplura or "Sucking Lice", and are to be distinguished from certain other animals which, owing to a superficial and not always very close resemblance, are also termed "Lice". Such are the "Bird Lice" (Mallophaga); "Bark Lice", "Book Lice", "Dust Lice", "Plant Lice", and "Louse Flies". All of these are insects and in addition there are "Fish Lice" and "Wood Lice", which are Crustacea. Sucking Lice occur on other mammals both wild and domestic and even on marine forms such as the walrus and seals. They are never found on birds.

The Biting or Chewing Lice (the Mallophaga) found on birds and mammals are probably related to the true Lice, but differ from them in having biting as opposed to highly modified piercing mouth-parts. They feed on pieces of the hair or feathers of their host and in some cases partly or entirely on blood or sebaceous matter. They are found on all birds and many of the mammals, including dogs, sheep and horses. On man their presence is entirely accidental.

An important characteristic of both Sucking and Chewing Lice is that they may show marked host specificity, that is to say any given species of louse may be found on only one species or a small group of species of host except by accident. Related hosts are usually infested by related species of lice. Thus the lice found on man for example have for their only close relatives the lice of the chimpanzee and gorilla.

Lice normally pass their entire lives on their host and if accidentally detached at once endeavour to regain their place on their original host or on one of the same species. It is thus that louse infestation spreads in a community. In this respect Lice and Scabies mites resemble one another in habit; they differ from Fleas, Bed-Bugs and Ticks which infest the places and premises inhabited by people but remain on their bodies only for the purpose of feeding. It is important to remember this point when planning methods of combating the different kinds of vermin because lice, to survive, must remain on their hosts except for very brief periods, whereas fleas and bed-bugs can remain alive for long periods in places uninhabited by man.

LICE (ANOPLURA) AFFECTING MAN

The species of lice found on man do not occur on other animals except occasionally on menagerie monkeys that have received an accidental infestation from human beings.

Man is infested by three kinds of lice commonly called Head, Body and Pubic Lice. The Head and Body Lice (Fig. 1) are regarded as varieties of a single species; the Pubic Louse (Fig. 2) is a separate species and of a different genus. The common names are derived from the parts of the human body that the lice normally affect. However, while the Body Louse is never found on the head, the Head Louse may be found on the body, including the pubic region. The Pubic Louse may be found on other parts of the body including the head.

The Head Louse and the Body Louse are commonly known as varieties *capitis* de Geer and *corporis* de Geer respectively of the species *Pediculus humanus* Linnaeus[1]. In the following pages the term "Louse" covers both these varieties unless one or other is indicated. The Pubic Louse is referred to as such; its scientific name is *Phthirus pubis* Linnaeus.

THE HEAD LOUSE AND BODY LOUSE

(*Pediculus humanus* var. *capitis* de Geer and *Pediculus humanus* var. *corporis*[1] de Geer)

A typical Head Louse and a typical Body Louse can be distinguished from each other by various characters. These differentiating characters are, however, all matters of degree and there are no hard and fast criteria to separate the two varieties. Thus the Head Louse is usually darker in colour, tougher in texture, smaller in size, with the segments of the antennae less attenuated, the abdominal clefts more deeply incised and the thorax narrower anteriorly than in the typical Body Louse. Head and Body Lice can be interbred and such matings produce fertile offspring which may show characteristics intermediate between those of the parents. In heavy infestations, some Head Lice may be found infesting the body; but Body Lice are never found infesting the hair of the head. These facts all point

---

[1] If these two varieties are regarded as sub-species, as they are by some authorities, then the one here called *Pediculus humanus* var. *corporis* would, in accordance with the International Rules of Zoological Nomenclature, have to be known as *Pediculus humanus humanus*, and the other as *Pediculus humanus capitis*.

to the conclusion that, in spite of the marked differences of their distinctive forms, Head and Body Lice are varieties of one species and that examples showing intermediate grades may occur. The Head Louse is regarded as the ancestral type and the Body Louse as a form that has been modified from it and adapted to life on the comparatively hairless human body, where it finds shelter in the clothing with which man has learnt to compensate his nudity.

Species of lice of the genus *Pediculus* found on the bodies or heads

FIG. 2. The Pubic Louse, *Phthirus pubis* Linnaeus. ×36.
(Photographed from a model in the Insect Gallery) [BMNH Neg. No. 50509].

of different races of man have been described by entomologists as different from the ordinary Head and Body Louse. Critical examination by competent authorities has, however, shown that all these may be considered to be *Pediculus humanus* though some may be given status as varieties of a standing similar to var. *capitis* and var. *corporis*.

Lice are not restricted to any particular areas of the globe nor are any of the races of mankind immune from attack or particularly subject to infestation. Man, thanks to the sweat glands in his skin and his own ingenuity in covering up his person with a variety of apparel, manages to keep the conditions of temperature and humidity next his skin fairly equable, regardless of the climate in which he is living, thus minimising for the louse the effects of climate. One may therefore expect to find lice wherever man makes his abode and neglects his standards of hygiene to a degree that is favourable to their multiplication.

THE PUBIC LOUSE

(*Phthirus pubis* Linnaeus)

The use of the colloquial term "Crab" as applied to lice refers to this species (Fig. 2). It is smaller than *Pediculus humanus* but the body is much broader in proportion to its length and the legs have the appearance of being attached to the edge of the somewhat flattened body rather than to the underside. Its movements are slow and deliberate and are made with a somewhat crab-like motion. As its common name implies, it is usually found infesting hairs of the pubic and peri-anal regions of both sexes. Like *Pediculus humanus* it is world-wide in its distribution, but much rarer, so that infestations of *Phthirus pubis* are less frequently seen.

THE LIFE HISTORY OF LICE

HEAD AND BODY LICE

It is convenient to describe the life histories of Head and Body Lice in the same section, since comparative studies have shown that their basic biological characters are closely similar. Thus, both adults and young of both sexes feed solely on blood and the methods of feeding and reproduction are identical in the two races. Furthermore, their rates of development, under various conditions, are the same. In fact, apart from certain contrasts in behaviour, the main difference is that Body Lice are generally more robust. This may be illustrated by longer resistance to starvation (about 85 hours at 23°C. or 73°F., as compared to 55 hours for the Head Louse). Body Lice are also more tolerant of rearing in captivity than Head Lice, which suffer rather high mortality when reared in gauze-topped cages. The greater adaptability of Body Lice is not surprising in view of the fact that its normal environment is rather less favourable than that of the Head Louse. The latter lives among the hairs of the head, a position which is exceptionally equable climatically, and the proximity of the scalp offers opportunity of feeding at any time of day or night. The Body Louse lives among the undergarments (especially those next to the skin) and the climatic conditions here are, indeed, far more uniform than those experienced by most other insects. However, garments may be removed at night (though they seldom are, by people chronically infested by body lice) and this possibility, as well as the movement of clothing during the host's period of activity, curtails opportunities of feeding. The method of feeding by lice is as follows: their mouth-parts are in the form of fine stylets which, except when feeding, are withdrawn within the head. When food is required, the skin of the host is pierced by means of the stylets which are thrust out and then gradually pushed farther and farther into the flesh. At the same time a salivary secretion, which prevents coagulation of the host's blood, is injected into the wound and enables the insect to suck it up without clots forming in the delicate mouth-parts. The adults can take up to one milligram of blood at a time. It is probable, however, as adults cannot live long without feeding, that they prefer to take smaller quantities at frequent intervals. At temperatures at which they remain capable of active movement they survive starvation, after a meal, for only a few days. At lower temperatures they survive for about a week after feeding, but it must be remembered that at these lower temperatures they are not capable of active movement. When a louse takes a very large meal, blood may sometimes be seen flowing from its anus. At other times over-eating results in a rupture of the gut and allows the blood to mingle with the body fluids; this causes the death of the louse within a comparatively short time.

Adults are capable of fairly rapid active movement. They have been recorded as moving at a rate of 9 inches per minute; but they never move in straight lines. Experiments have shown that a louse-free person lying in the same bed as a lousy person will begin to complain of lice in 5–6 hours.

Pairing between the adults may take place within the first ten hours of adult life and is repeated frequently throughout the remainder of their lives. In pairing, the male approaches the female from behind and works his way beneath her abdomen. He grasps her hind legs with his fore-legs and, when the genital organs at the tips of their abdomens have coupled, the female rears up into a vertical position lifting the male off the ground. They then return to the horizontal position and may remain united for a period varying from half an hour to several hours.

The female is oviparous. Under experimental conditions lice kept in contact with the human body will commence to lay eggs in from 24–36 hours after moulting for the last time. When periodically removed from the skin, as Body Lice would be in clothing removed at nights, they begin to lay eggs in about four days after the commencement of adult life. Parthenogenesis does not occur.

A female Body Louse may lay up to a dozen eggs per day under optimum conditions. She probably lives for a maximum of about a month, so that total egg numbers of 300 or more have been recorded.

The figures for Head Lice are considerably lower; but one cannot say whether this is due to their being more handicapped by the unnatural rearing in cages necessary for such studies, or whether they are actually less prolific in nature. In any case, "optimum conditions" imply that the lice are not interfered with; whereas they are very likely to be injured or killed by the scratching and other reactions of their host. It is likely that the proportion of females which survive to die of old age is quite small. Body Lice have the added hazard of being isolated from their host if clothing is removed. They will not lay eggs if starved nor if the temperature falls below about 20°C. (68°F.).

The eggs or "nits" (Fig. 3) are laid singly and measure about

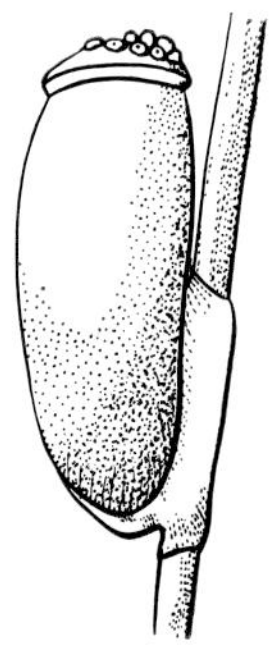

FIG. 3. The Egg of *Pediculus humanus* attached to a human hair. ×64.

0·8 mm. in length by 0·3 mm. across. The egg has a cap or operculum at one end and is attached to the host's hairs or the fibres of clothing by cement secreted by the female at the time of laying the egg. The eggs of the Head Louse are usually laid singly near the base of the hairs, but in severe infestations there may be more than one nit on a single hair. The situation most favoured by the Body Louse for her egg-laying operations is in the folds of cloth at the seams, especially of the undergarments.

The young nymph hatches after an incubation period of from six days to a fortnight. The length of incubation depends on the temperature and on whether the eggs have been kept continuously at that temperature or periodically chilled, as would be those laid on clothing taken off at night. If maintained continuously at the temperature prevailing on the surface of the human body (30°C., 86°F.), eggs hatch in 8 or 9 days. If kept continuously at 40°C. (104°F.) eggs die in a couple of days. Maintained at temperatures below 20°C. (68°F.) eggs develop slowly, but the nymph cannot emerge and dies in the egg-shell; but if such eggs are removed to a warmer environment they may hatch, provided the exposure to cold has not exceeded 14 days. Exposure to a temperature of 8°C. (46·4°F.) kills in 7 days.

When ready to hatch, the nymph pushes off the operculum and forces its way out of the egg. As it comes out it casts its skin, which it leaves behind within the empty egg-shell.

The young nymph is about 1·0 mm. long. In general appearance it resembles its parents and its mode of life is similar. It takes blood and will starve to death in about the same time as an adult. The nymph moults its skin three times before reaching maturity. Under experimental conditions it has been found that nymphs kept in continuous contact with the human body become adult on about the eighth day after hatching. When kept on the body by day and removed at night, as they would be in the case of nymphs living in clothing, the period was extended to 13–16 days.

The empty egg-shells left behind by the nymphs have a characteristic appearance. They are rather "papery" and may have small dents in their surface, and the operculum is lost or hangs loose. They remain attached to the hairs or clothing fibres for a long time after the nymphs have hatched and in the case of the Head Louse are carried by the growth of the hair away from the base where they were originally deposited. The appearance of these empty shells is important since they may remain attached to the hair and clothes after delousing. The presence of nits, is, therefore, not a definite symptom of present lousiness unless it is verified that the nits are alive and not just empty egg-shells.

The activity of lice is vitally affected by the temperature of their surroundings. They do not survive abnormally high temperatures. For instance 50°C. (122°F.) will kill the egg in 30 minutes, and a rather lower temperature suffices to kill the nymph or adult in the same period. Rather higher temperatures are more rapidly fatal.

The temperature of the surface of the body (about 30°C. or 86°F.) would appear to be the best suited to lice and, as already mentioned above, they leave a person with a fever or a dead body that is becoming cold. They are capable of activity at ordinary comfortable room temperatures, but their activity decreases as the temperature becomes lower, though the precise point at which it ceases is not known. Temperatures round about freezing-point are fatal if the exposure is prolonged, but lice can withstand them for considerable periods if warmed up and fed every day. These lower fatal temperatures are not, however, of great importance, since it is not practical to employ low temperature for disinfestation.

The Pubic Louse

Information about the Pubic Louse is scanty. The eggs, smaller than those of the Head or Body Louse, are attached to the hairs of the host by a relatively larger amount of cement and are never found on the clothing. The nymph hatches from the egg after an incubation period of about 7–8 days. There are three nymphal stages and the nymph becomes adult about 13–17 days after hatching. The adult probably lives not more than a month, during which time the female seems to lay fewer eggs than does the Head or Body Louse. It is much slower in its movements, does not move about so much and it seems to be unable to survive for longer than a day when removed from its host.

The Pubic Louse is most commonly found on the hairs of the pubic and peri-anal regions of both sexes. It can, however, also occur on the other parts of the body where there is a growth of hair, even on the eyelashes and eyebrows. Only rarely does it occur on the scalp, but this may be explained by the fact that its claws appear to be of a size best suited to holding on to the rather coarser hairs of other parts of the body. Infants have been found to harbour Pubic Lice on the scalp, and this fact, along with the existence of infestations of the eyelashes with a complete absence of infestation of the pubic regions in the adults of a community, shows that direct head to head transmission does occasionally take place. From these facts it is evident that transmission from one person to another may be due to close but casual contact. The gravest injustice is done by assuming that transmission only takes place during sexual contact, though that is no doubt common.

The Pubic Louse tends to spread amongst people living together under bad conditions. It is rarer than the Head or Body Louse but, like it, it seems to occur amongst human beings wherever low standards of hygiene exist.

LICE IN RELATION TO MAN

How Infestation Arises

Lice can only be acquired, directly or indirectly, from an infested person; and the likelihood of this happening depends on (i) the general prevalence of infestation in the community and (ii) the intimacy of contact. Both matters require separate consideration for Body and Head Lice.

(i) Under modern civilized conditions, Body Louse infestations are very rare and are virtually restricted to people of obviously low standards of hygiene, such as vagrants, who rarely change their underwear and commonly sleep in their clothes. (Regular laundering of undergarments makes it impossible for Body Lice to survive.) Such people tend to associate together, either sleeping "rough" or in common lodging houses with low standards of hygiene; so that lice tend to spread among them. The authorities may have to cope with the problem of disinfestation when such persons come to public assistance hostels or to prison. For the great majority of the population in modern Britain, the chances of becoming infested with Body Lice is remote.

Body lice are still quite prevalent in some communities, with low hygienic standards, or in the absence of civilized amenities. Furthermore, they can become widely dispersed among any large groups of people unable to wash and change their clothing; for example, due to the upheavals of war or large scale disasters. Until the introduction of DDT in the last war, combat troops and prisoners of war were commonly lousy for much of the time.

Head Lice are unfortunately much more prevalent than Body Lice, even under civilized conditions. Public health authorities were dismayed at the extent of infestation revealed during the last war. Records of admission to fever hospitals (chosen to give an unbiassed sample) showed that nearly 50% of children up to 5 years of age, in certain English cities, were infested with Head Lice. Girls of school age up to 14 years old were almost as lousy, but boys of school age were much less verminous. In rural areas, the incidence of lousiness among similar children was only 5%. Figures given by public health authorities for lousiness among children are generally much lower than this; but it must be remembered that, with the best intentions, school authorities cannot carry out as thorough an examination as hospitals.

From more recent evidence, it is clear that there has been a decline in the prevalence of Head Lice, but the insect is far from being exterminated. Records of the Public Health Department of one British city provide the following figures:

		1940	1943	1953	1963
Boys	Lice %	3·7	2·6	1·6	0·1
	Nits %	17	6·8	6·3	2·7
Girls	Lice %	15	13	5·0	0·1
	Nits %	40	27	21	11

It is therefore not entirely uncommon for children to pick up a few Head Lice at school or play, though a conscientious parent will make sure that the infestation does not persist. Unfortunately there seems to be a small proportion of families, some members of which are nearly always infested; and these form a small reservoir of lice which may spread to others. This problem is very difficult to deal with, short of draconian measures.

(ii) Normal transmission of Lice probably occurs under conditions involving close proximity or contact, as, for example, in crowded sleeping quarters or on occasion when people are forced to live crowded together. It is easy to understand, therefore, why widespread Body Louse infestation is often a result of war or disasters, in groups of prisoners or refugees. As regards the Head Louse or Crab Louse, the intimate contacts of family life are ideal for transmission.

Accidental transmission may also occur in a variety of ways, as, for example, by direct contact in crowded public resorts or conveyances, or indirectly by means of cushions, upholstery or seats. The use in common of bedding, towels, clothing or hair brushes and combs by a number of people may also serve to spread lice from one person to another. Stray hairs and combings may convey lice or their eggs and it is likely that this is onc mcans of transmission of Pubic Lice. Another possible source of infestation is the seats of ill-kept public water closets.

While all these possibilities exist, it must be borne in mind that all forms of lice fare badly away from their normal environment and soon starve to death. Accordingly, the risks of transmission decrease steadily, from the intimate contacts of family life, to the close association of children at play and finally to the remote risks of infestation via fomites (furniture and other articles used by the infested person).

Effects of the Habits of the Host

Three aspects of human behaviour may be considered in relation to louse infestation. Head Lice will be affected by the state of the hair and Body Lice by that of the clothing. Where conditions are such that an infestation has developed, both forms of lice will be affected by the reactions of the infested person.

Head Lice and the Hair

Such customs as shaving the head will prevent infestation of the scalp with Head Lice and the close cropping of the hair customary amongst soldiers serves the same purpose. Modern standards of hygiene in respect of the hair, and the frequent combings and brushings that it receives in circles where any attention is paid to cleanliness, must prevent many infestations, though perhaps not carried out deliberately for this end. Lice may be expected to thrive when once they get a foothold on a head where the hair is considered to be exempt from washing and thorough combing, a state of affairs that, curiously enough, may arise through attempts to prolong the "life" of a permanent wave, amongst persons to whom the acquisition of such a luxury represents a very considerable outlay of money.

Persons with an abundant growth of hair on the scalp are more liable to infestations of Head Lice than those with short hair. The sex of the host has no influence on the liability to infestation except in so far as custom or fashion may dictate that one or the other sex shall wear the hair short or long.

Although Head Lice tend to be discouraged by normal brushing, combing and hair washing, one could not guarantee that such operations would exterminate an infestation, unless very thoroughly done. For this reason, it is perhaps not surprising that among the poorer children in an industrial city, it was shown that the liability to infestation increased with the size of family (at least up to 5). No doubt this reflects the declining opportunities for close attention by the more harassed mothers.

Body Lice and Clothing

It has frequently been alleged that lice show a seasonal variation in their incidence. There are, however, few figures available from which any safe conclusions can be drawn with regard to this suggestion. Such fluctuations as have been said to exist are more probably due to seasonal changes in the dress of the people rather than to the direct effects of seasonal changes on the insects themselves.

Clothing, as might be expected, has a considerable effect on populations of lice on the body. Man uses clothes to avoid violent fluctuations of temperature on the surface of his body. This is ideal from the louse's point of view. When clothing is frequently changed and efficiently laundered, dry-cleaned or even laid aside unused for a long period, lice and their eggs will not survive; on the other hand clothing which is continuously worn day and night provides ideal conditions. Bathing, moreover, loses its value in the case of lousy persons if the old clothes carrying the lice and nits are put on again without being freed from the pests. Amongst certain peoples with low hygienic standards, the amount of clothing worn is often a mark of wealth, power or dignity. Thus amongst certain African tribes the adult males, with their voluminous robes will be found to be much lousier than the half-naked adolescents. An example taken from history concerns that great Church dignitary St. Thomas à Becket whose underwear, made of hair-cloth, is described by a contemporary chronicler as being covered with innumerable lice which "boiled over" as the corpse was stripped of its heavy vestments on the morning after his murder.

Host Reactions to Infestation

Some individuals are more sensitive to the bites of lice than others. Individuals who are insensitive to them may be indifferent and thus act as possible centres for the dissemination of the vermin. A similar indifference may arise in a normally sensitive person who has had to endure an infestation for a considerable time without means of getting rid of it.

The natural reaction of a sensitive person to louse bites is to scratch the part affected. Experience does not suggest that this results in the violent death of any large numbers of lice. The quality, quantity and design of clothing may, however, prevent indulgence in efficient scratching of the body, while consideration for elaborate coiffures may deter both savage man and civilized woman from scratching the scalp. In addition, the social taboo that exists in many circles against scratching also tends to lower the efficiency of this natural and moderately effective response of human beings to the attacks.

Individuals who are habitually lousy usually indulge in a deliberate primitive delousing operation by hand picking. This is, however, directed only against such individual lice as can be seen with the naked eye. Many nits will almost invariably be passed over and perhaps some of the nymphs, so that this operation would only by a fortunate accident cause the extermination of an infestation. There are, of course, chronically lousy individuals who have ceased to worry about their condition and who take no such steps to deal with their parasites.

Although the scratching and hand picking operations of lousy people would not be expected to terminate infestation, the following facts are of some significance. If one counts the lice in a garment or on a lousy head the number found is commonly low. About ten to twenty Head or Body Lice per person is usual, but sometimes a hundred are found in a single head of hair, and rare cases are known in which over a thousand have been recovered.

The fact that the most common infestation is only of 10–20 insects is of considerable interest. It means that there is very large mortality due, no doubt, to the host's reactions to the irritation caused by their bites. It seems very probable that habitually lousy individuals have each a characteristic normal louse population and that a 20-louse-man would only become a 200-louse man as the result of some very radical change in his habits enforced by some outside influence.

LICE AND DISEASE

Direct Irritation

Head and Body Lice. When the louse bites, it is the reaction of the host's tissues to the injected saliva that causes the symptoms of irritation associated with louse bites. In most cases the bite results in the formation of a small red papule with a tiny clot in the centre at the actual point of penetration. The symptoms may persist for several days but the length of time and the degree of irritation vary in different individuals. Exceptionally, louse bites may give rise to a general or local urticaria and itching. After being exposed to the attacks of lice for a prolonged period some people become immune to their bites and show no reactions. Individuals who have been habitually lousy for a long time may eventually exhibit a thickening and discoloration of the skin known as "Vagabond's Disease".

Pubic Louse. The bite of the Pubic Louse gives rise to a similar irritation. A characteristic "blue spot" from 0·2 to 3 cm. across, may result from the bite; these spots are not found after the bites of the Head or Body Louse.

Skin Infections

Lice themselves do not cause skin troubles. Scratching, however, which is the natural reaction to their attacks, combined with unhygienic conditions, may result in abrasions of the skin being infected and the production of impetigo or a dermatitis. There are cases on record in which the exudates from an infested eczematous scalp have formed a crust with the matted hair, beneath which the lice throve in large numbers. Cases of blood-poisoning and death resulting therefrom have been attributed to heavy infestations of lice, but here again it is the infection of a scratch that is the real cause.

Louse-borne Diseases

Insects may be concerned with the transmission of organisms which cause diseases in man in two different ways. They may act merely as mechanical conveyors of the organisms from one person to another, in which case there may be many other ways in which the disease can be transmitted. In the second class the organisms cannot get from one person to another without the intervention of an insect. In this latter class the pathogenic organisms are taken up by a blood-sucking insect from the body of a diseased host and, after a period in the insect, during which they may undergo considerable development and multiplication, perhaps accompanied by a radical change in form, they are passed into a new host by some act of the insect.

Experiments have been carried out which indicate definitely that *Pediculus humanus* can transmit impetigo. This is a case of purely mechanical transmission, the organisms causing the disease being carried by the louse from a diseased host to a clean host.

In the second class there are two well-known diseases that are transmitted by *Pediculus humanus* and under ordinary circumstances cannot be transferred from one man to another in any other way. These diseases are Exanthematous Typhus and Louse-borne Relapsing Fever.

Phthirus pubis is not known to transmit any diseases of the second class.

Typhus.[1] The cause of Typhus is an organism called *Rickettsia prowazeki* da Rocha Lima. These organisms are extremely small.

A typhus patient is capable of infecting a louse quite early in the course of the disease and remains so till about the tenth day after the commencement of symptoms. A louse biting a patient during the infective period sucks up large numbers of the Rickettsia in the blood. Once in the louse's gut the Rickettsia multiplies rapidly and penetrates the layers of cells that form the sides of the louse's gut. The Rickettsia that get into the cells proceed to multiply rapidly, the others seem to die in the gut. When multiplication has taken place the cells are much distended and eventually burst into the cavity of the gut. When this has happened the Rickettsia is to be found in large numbers in the louse's faeces. The presence of the Rickettsia in the gut and their effect upon the cells of its walls so upset the louse's constitution that it eventually dies, usually about ten days after the rupture of the cells. During the whole period following the rupture of the cells of the gut wall and till its death, the louse continues to produce faeces containing Rickettsia.

Man acquires infection when he rubs some of these faeces containing the Rickettsia into some abrasion or scratch on his skin. The infected faeces may come either from a live louse that has recently been acquired from a lousy typhus case or possibly from faeces in bedding and clothing since it has been shown that Rickettsia remains capable of causing infection for a considerable time after having been passed in the louse's faeces. It also seems possible that infection may

[1] Exanthematous Typhus Fever, also known as Jail Fever, Camp Fever and Ship Fever, must on no account be confused with Typhoid Fever, which is a totally different disease due to a Bacillus and not a Rickettsia.

be picked up from the faeces falling on the conjunctiva or being taken into the lungs by inhalation.

Since typhus can be spread from man to man only by means of the Rickettsia in the faeces of lice it is obvious that the disease cannot exist in the absence of these vermin. There are many lice in England, but there have been no deaths from typhus since 1919. Infestation by lice was the rule rather than the exception at the front during the war of 1914–18, but there were no cases of typhus on the western front. There are regions in the world where typhus is always present and where a deterioration of living conditions, with the resulting increase in infestation by lice, leads almost automatically to an epidemic of typhus, as in the Balkans during the same war. During the war of 1939–45 there was little trouble from widespread typhus epidemics; this was, in fact, a direct outcome of the great care taken to prevent infestation by lice.

It is therefore obvious that where typhus is endemic the importance of keeping down the infestations of lice cannot be over-emphasised. At the same time our present stage of knowledge of typhus is not sufficient to enable us to foretell the precise conditions under which typhus will appear in epidemic form and in consequence lousiness should always be combated, especially when it is the result of a recently lowered standard of living general throughout the population. Delousing was recognised as a means of combating typhus as long ago as 1876, but the mode of transmission was then unknown and the actual transmission through the faeces only elucidated a quarter of a century ago.

Exanthematous typhus is essentially a disease of temperate climates. It is found in Eastern Europe, Russia, the Balkans, the Mediterranean region, Abyssinia, South Africa, China and the non-tropical parts of Asia. It is absent from the plains of India, but has been recorded from most but not all tropical countries. It used to occur in western Europe and there were over three thousand deaths from it annually in England and Wales till about 1870. Since that date it has steadily decreased till it disappeared in 1919. It is absent from the U.S.A. except for a mild form known as Brill's Disease. Zinsser in his book *Rats, Lice and History* (Routledge, London, 1935) has given a vivid picture in non-technical language of the important role that typhus epidemics have played in the history of mankind and a perusal of it will demonstrate the vital importance of controlling the louse.

Relapsing Fever. Man is affected by several so-called Relapsing Fevers. These are all characterised by a fever which recurs at intervals in much the same way as in malaria, for which relapsing fever has very frequently been mistaken. The relapsing fevers are due to infection by different species of a group of minute organisms called

Spirochaetes. The Spirochaete (*Spirochaeta recurrentis* Lebert) which causes Louse-borne Relapsing Fever is transmitted from man to man by *Pediculus humanus*.

The blood of a victim of louse-borne relapsing fever contains numerous Spirochaetes and when a louse bites such a person it takes up large numbers of them with the blood it imbibes. Many of these are digested in the louse's gut but some penetrate the cells forming the wall of the gut. What exactly happens to the latter is not known but, after a period which may be quite short or as long as six days, Spirochaetes reappear in large numbers in the body fluids of the louse. These Spirochaetes remain in the body cavity and do not make their way to any of the organs of the louse, such as the salivary glands or the gut, from which they could possibly escape. In fact the only way in which they can escape from the louse is through the latter being injured or killed in such a way as to allow of an escape of the body fluids.

Man, when infested or troubled with lice, now and then succeeds, in the course of his scratching, in crushing a louse or injuring it in such a way that some of its body fluids escape on to his skin. If the skin happens to have abrasions at that point Relapsing Fever Spirochaetes may find their way into his system through them, multiply, and cause the symptoms of the disease to appear.

From what has been said above it will be seen that the practice of "popping" lice with the finger-nails is a highly dangerous practice if relapsing fever is prevalent. The even less civilised method of killing individual lice, caught in the course of personal delousing operations, by biting them, is to be deprecated on the same grounds.

Louse-borne relapsing fever is widespread. It appears, however, to be absent from western Europe (including the British Isles), U.S.A., Malaya, Australasia and Tropical Africa south of the country bordering the southern side of the Sahara Desert.

PREVENTION AND TREATMENT OF LOUSINESS

General Considerations

Lice are sometimes prevalent among troops, prisoners or refugees, but, with the advent of powerful new insecticides, eradication of lice from such people is comparatively simple, because they are subject to discipline, or control. But the reduction of levels of infestation in large civilian communities may present problems apart from the efficiency of the delousing measures. A very great deal depends on the mental attitude of individuals as well as on their standards of hygiene.

In circumstances where louse-borne diseases are prevalent there is little or no difficulty in securing co-operation of the people. The use of general delousing measures among a large population to quell a typhus epidemic was undertaken first in Naples in 1943 and has been successfully accomplished in other places subsequently. It is, however, another matter to eradicate lice when there is no obvious threat of disease. The essential difficulty is to convince an inert or reactionary section of the community of the necessity for exterminating lice. In backward communities there may be a high proportion of such people; in advanced societies there may be only a few, but they are a chronic nuisance.

It is important for all people to understand clearly that lousiness is preventable; that it is a condition that can and should be stamped out; that there is no more social disgrace attached to becoming infested than to catching measles, and finally that the methods of overcoming lice are simple.

Quite apart from the reluctance to admit to harbouring an infestation of lice lest it be considered a mark of special disgrace, certain popular beliefs stand in the way of educative measures for louse eradication. Some sections of the community actually hold the belief that the presence of lice is a sign of health; this takes the form of such tags as "You're not healthy, you've no nits." A similar popular misconception is that lice arise by spontaneous generation from dirt. Until people understand the fundamental fact that lice arise only from other lice it is difficult to help them to overcome the pests. Another belief that may cause difficulties is the theory held by many people that certain individuals are "breeders", it being thought that lice proceed out of their skins and that nothing can ever be done to stop this process. It is also popularly but unjustly believed that infestations of pubic lice can only be picked up by sexual contact and that an infestation may be considered evidence of irregularities in this respect. While pubic lice may spread in this way it is known that they are also spread as nits attached to loosened hairs that adhere to towels, lavatory seats, etc., and in other ways.

Erroneous and obstructive ideas about lice can be countered by education and propaganda. Children should be taught at school the essential facts of hygiene in relation to the eradication of lice, for there is evidence that the incidence of lice amongst school children is very much higher than is usually supposed (see p. 9). To introduce more frequent and thorough inspections would be difficult, but the possibility of educating children to recognise and admit infestations when acquired should be explored. Adults can be reached by means of posters and other forms of publicity.

The salient fact to be presented by such educational propaganda is that cleanliness of the person and of the clothing is the primary defence against lice. Lice can be picked up by any one; but they are unlikely to survive long on people who brush and comb their hair frequently and thoroughly, who change their underwear at least weekly and have it properly washed. Modern laundering and dry-cleaning practices are fatal to lice; and though woollen fabrics are sometimes washed at temperatures below those fatal to lice, ironing and other treatments make it hard for the insects to survive.

Washing the head and body, though highly commendable, cannot be guaranteed to destroy lice because water temperatures tolerated by the human skin are not lethal to them. This may be a reason why quite cleanly children can develop head louse infestations. It emphasises the importance of thoroughly combing and brushing the hair daily: and this is much easier for girls and young women if they adopt short, free hair styles.

While individual cleanliness is the most important weapon against infestation there are certain obvious precautions which reduce the chances of picking up stray lice. The use of communal clothing and toilet articles is to be avoided. For example, children should not share brush or comb with others at school, or exchange hats. Close contact with persons likely to be harbouring lice is naturally to be avoided, as is the use of lavatories that are not above suspicion.

No confidence should be placed in fabrics said to repel lice and no satisfactory repellent is known which will keep them away. On the other hand, DDT powder dusted on the underwear next to the skin (as described later) will kill lice as they are acquired and is thus, for practical purposes, equivalent to a repellent. This practice, which is also effective against fleas, may be useful for travellers in certain uncivilised parts of the world.

Disinfestation Methods

Apart from certain resistant strains of Body Lice, which have developed partial immunity to DDT, it seems that all types of human lice are susceptible to modern insecticides, fumigants or physical control measures. But the particular circumstances of various types of infestation require different criteria for deciding on the most satisfactory control measures. (Thus, in some cases, protection from reinfestation is more essential than rapid action; while the importance of using cosmetically acceptable treatments depends on whether or not there is a threat of louse borne disease.) For these reasons, the control of Body Lice under conditions of widespread lousiness is discussed separately from the problem of dealing with relatively small numbers of lousy individuals in a clean environment.

Control of Head Lice is described in a third section and control of Pubic Lice in a fourth.

Control of widespread infestation by Body Lice

In circumstances of general lousiness, it is essential to reduce the infestations as quickly and widely as possible. The measure must be quick, safe and effective, and provide some protection from reinfestation, which would otherwise occur. Owing to the danger of disease, however, people will accept treatments which would be resented in other circumstances.

The ideal way to deal with this emergency is to apply insecticidal dust to the underwear of every member of the community, as soon as possible. If the lice are not resistant to DDT, a dust containing 10% DDT is the insecticide of choice. DDT-resistant lice may be killed by dusts containing 1% malathion or 1% *gamma* BHC.

The dust can be applied by individuals to the underwear, using about one ounce (30 gm.) per person. With sifter-top cans, the powder is applied evenly over the inner surface of the garments next to the skin, with special attention to the seams. The seams and folds of other garments and also the socks should be similarly treated.

For the mass treatment of large groups of people the clothing need not be removed and hand-operated dusters or motor-driven air compressors with as many as 10 duster heads can be used. About 50 gm (say, 2 ozs) of powder is shaken or blown into the clothing through the neck openings, up the sleeves and from all sides of the loosened waist of trousers. In delousing women, an extra quantity may be introduced down the neck of the dress and application at the waist-line omitted. The socks, head covering, the inner surfaces of extra garments, and the bedding should also be treated.

Disinfestation of individuals with Body Lice, entering a clean environment

Under civilized conditions, the usual problem of Body Louse control is to deal with relatively small numbers of infested people who come to public assistance hostels, charitable institutions or prison. In such situations DDT dusting is unsuitable, since the obvious signs of being heavily dusted will generally be strongly resented. Furthermore, the toxic action of DDT is slow, so that intoxicated lice may be seen crawling on treated people for some time after treatment. Since there is no danger of reinfestation, a rapid treatment giving no residual protection is adequate; for example by fumigation, or heat treatment. While individuals are having their clothing disinfested, the stray lice on the body should be removed by shaving off body hair and bathing with plenty of soap. This was the practice in the early type of delousing stations perfected at the end of the First World War. The procedure is described in detail in Professor Buxton's book, "The Louse" (1947).

Fumigation. Probably the easiest treatment is to fumigate the clothing (or bedding, where necessary) with ethyl formate in a metal bin or a plastic bag. A dose of 2 oz/cu.ft. should be used for a short (1 hr.) exposure; but for 5 hrs. or longer, $\frac{1}{2}$ oz/cu.ft. is adequate. The short treatment could be given while the infested person took a bath, while the longer fumigation could be used to treat clothing overnight. The cost is low and the time to load and unload the clothing takes only a few minutes. On their removal, there is only a mild smell of ethyl formate, which disperses rapidly. The inflammability and toxicity of ethyl formate is about the same as benzene, so that it can be handled by any reasonably sensible person.

Heat treatment. An alternative to fumigation is heat treatment. Dry air is a poor conductor of heat so that a temperature well above that sufficient to kill exposed lice or eggs (see p. 7) is required, to penetrate into clothing or bedding. Garments should be kept in air over 70°C (158°F.) for an hour, taking precautions against scorching where necessary.

The large steam disinfestors used in hospitals to disinfect bedding can be employed for de-lousing. But the very high temperatures, required to destroy bacteria, are not necessary to kill insects and the whole procedure is rather cumbrous for this purpose.

Control of Head Lice

Shaving of the head is an old-fashioned and very thorough remedy, but unnecessarily drastic and should not be used, except in severe cases, complicated by skin infections, which cannot be treated without removing the hair.

Head infestations can be dealt with either by the use of insecticides or by combing with a special fine-toothed comb.[1] Either method, however, used alone has certain disadvantages. No known insecticide removes nits though some kill them. Since it is not easy to distinguish dead nits from live ones, it is best to remove them. Combing with a specially fine metal comb will effectively remove all lice and nits, providing the combing is carefully and conscientiously done. Special attention must be paid to hair at the back of the head

[1] These special combs can be obtained from Messrs. Sacker's Hygienic Comb Co. Ltd., 8 Mount View Road, London, N.4.

and behind the ears and the comb worked well down to the base of the hairs so that neither nits nor lice escape. Combing alone, however, apart from being slow and laborious, gives no protection from reinfestation. Incidentally, it may be pointed out here that the belief that vinegar (acetic acid) loosens dead nits and allows them to be easily combed out, is false. No substance is known which will dissolve the cement of the louse egg without damage to the human hair. However, damping or slightly oiling the hair may lubricate it and allow the fine metal comb to pass through it more easily.

Probably the best treatment is a thorough combing followed by application of a persistent insecticide. Other members of the same family should apply some insecticide as a prophylactic, because there is always a considerable chance of infestation spreading through a small group of people living together. The hair should not be washed for a week after treatment, because the insecticide remains active for that time and will kill any young lice that hatch from nits that may have escaped destruction. After a week the hair should be washed and carefully inspected for lice, of which it should be quite free. However, a second dose of insecticide should be applied as a precaution. The following insecticides have been widely used and proved to be most reliable:

0·2% *gamma* BHC suspension
2·0% DDT emulsion

Both are applied in the same way, as follows: about 3 to 12 ml. (one to four teaspoonfuls), according to the amount of hair, are required. The chosen liquid is applied here and there to the roots of the hair, which is rubbed with the fingers to distribute the insecticide. Finally, an ordinary comb is used to tidy the hair.

DDT powder has been found to be effective against head lice, but since it gives the hair a chalky appearance, its use is usually resented by patients.

Control of Pubic Lice

As with head lice, the earlier treatments involved shaving or cutting close the infested hair; but this should not be essential now. The simplest and most satisfactory remedy is to apply one of the modern insecticides and repeat the treatment one week later to kill any young lice hatched from unaffected eggs. A dust containing 5% DDT or 0·5% *gamma* BHC or an emulsion containing 2% DDT or 0·2% *gamma* BHC would be effective. The centuries-old use of mercurial ointment cannot be recommended; its insecticidal value is low and there is some risk of mercurial poisoning.

In treating infestations of crab lice in the eyelashes, it is best to use a vaseline ointment containing pyrethrins and to remove the insects and nits individually with fine forceps, using cocaine on the conjunctiva if necessary.

Disinfestation by Public Authorities

The following empower (or assist) local authorities to deal with lousiness: Education Acts, 1921–1937; Public Health Act, 1936 (sections 83–86); Public Health (London) Act, 1936 (sections 122–127); Scabies Order, 1941. For details of the application of these Acts, see the Ministry of Health Memorandum on the Louse.

Most Public Authorities in urban areas maintain cleansing stations, generally in connection with Public Bath Houses. Many people, however, will prefer to take the trouble to delouse themselves rather than accept the facilities freely offered by the local authorities.

Acknowledgements

The author of the first edition of this booklet, Dr. J. Smart, was indebted to the late Professor P. A. Buxton for helpful suggestions and to Dr. J. Macleod, Mr. H. J. Crauford-Benson and Dr. K. Mellanby for additional information. The second and third editions were revised by Professor P. A. Buxton and Professor J. R. Busvine; and the latter was responsible for revising the present edition.

WHERE TO GET MORE INFORMATION ABOUT LICE

Bibliography

Professional entomologists, medical men, public health officers and others seeking further scientific information should refer to:

The Louse by P. A. Buxton (Arnold, London) 2nd ed. 164 pp., 1947.
A monograph on lice affecting man.

Insects and Hygiene by J. R. Busvine (Methuen, London) 2nd ed. 467 pp., 1966.
A chapter on lice, with additional information.

Recommended Methods for Vector Control. Appendix to 17th Rept. Expert Committee on Insecticides. World Health Organisation, Geneva, 1969. Up-to-date information on control.

Contribution towards a Monograph of the Sucking Lice. VIII by G. F. Ferris. (Stanford University Publication) Biol. Sci. 2. No. 8, 1935.
Details of systematic classification of lice.

Rats, Lice and History by H. Zinnser (Routledge, London) 1935.
Good popular account of relations of lice to man.